Biscuit, cookie and cracker manufacturing manuals

Manual 2
Biscuit doughs

Please personalise your copy with your name below

....

....

Biscuit, cookie and cracker manufacturing manuals

The other titles in this series are:

Manual 1
Ingredients
Types • Handling • Uses

Manual 3
Biscuit dough piece forming
Sheeting • Gauging • Cutting • Laminating • Rotary moulding • Extruding • Wire cutting • Depositing • Troubleshooting tips

Manual 4
Baking and cooling of biscuits
What happens in a baking oven • Types of oven • Post-oven processing • Cooling • Handling • Troubleshooting tips

Manual 5
Secondary processing in biscuit manufacturing
Chocolate enrobing • Moulding • Sandwich creaming • Icing • Application of jam • Marshmallow • Caramel • Troubleshooting tips

Manual 6
Biscuit packaging and storage
Packaging materials • Wrapping operations • Biscuit storage • Troubleshooting tips

Biscuit, cookie and cracker manufacturing manuals

Manual 2
Biscuit doughs

Types
Mixing
Conditioning
Handling
Troubleshooting tips

DUNCAN MANLEY

WOODHEAD PUBLISHING LIMITED
Cambridge England

Published by Woodhead Publishing Limited,
Abington Hall, Abington, Cambridge CB1 6AH, England

First published 1998

British Library Cataloguing in Publication Data
A catalogue record for this book is available from the British Library.

ISBN 1 85573 299 3

Designed by Geoff Green
Typeset by BookEns Ltd, Royston, Herts.
Printed by St Edmundsbury Press, Suffolk, England.

Contents

Preface

My text *Technology of Biscuits, Crackers and Cookies,* which was first published in 1983, with an enlarged new edition in 1991, has proved very successful and been welcomed by biscuit manufacturers worldwide. Why, then, consider producing separate manuals in the same field?

The idea started, I suppose, when my partner, Pam Chance, pointed out that, as a standard reference work, my book was both too detailed and expensive for the average plant operative to use in the course of his or her work.

Over the years, I have worked as a consultant in very many factories in many countries. Not all the operatives speak English, but those that do have explained that they often do not know the biscuit technology with which they are involved nor have a convenient source of information. They and their managers have particularly expressed the need for aids to troubleshooting.

Thus this manual was born. It is one of a sequence that covers the various parts of the biscuit-making and packaging process. It builds on *Technology of Biscuits, Crackers and Cookies,* but includes much new material. I have tried to give particular emphasis to process mechanisms and fault solving. I am sure that managers, trainers and operatives will find it useful both in training and as a reference source.

I hope that all who read and use it will find it as useful as I would like it to be. If you have any comments or contributions, I should be pleased to hear from you.

Duncan J R Manley
1998

1 Introduction

This manual is one of a series dealing with materials and manufacturing procedures for biscuits.

It describes, in general terms, what is involved in making biscuit doughs and the subsequent handling of this dough. The techniques used to form the dough into pieces for baking are dealt with in Manual 3, *Biscuit dough piece forming*, and details of the ingredients used are in Manual 1, *Ingredients*.

If you are a member of a manufacturing team you should know how to do your tasks and the **reasons** for doing things in a specified way. You should also understand the possible implications of not doing a task correctly or not communicating difficulties promptly. The different types of biscuit doughs are described. Critical aspects to watch for during their preparation and subsequent handling are also described. It is essential for the efficient running of the factory that doughs are made correctly since it is much better to prevent errors than to correct them later.

From time to time problems will arise. A competent operator will be able to recognise and correct faults quickly. Sections are included in this manual which describe common faults and problems and should therefore aid troubleshooting and problem solving.

If you work in a food factory you must accept some responsibilities. These to a greater or lesser extent will include:

1 Responsibility for the quality of the company's products if you are a member of a production team. By critical observation and knowing what to look for and expect, you could prevent a fault occurring during manufacturing.
2 Responsibility for the safety of consumers who will eat the products you help to make and pack. The procedures and precautions you should observe are described.
3 Responsibility for the machines and equipment with which you

are required to work. The procedures and precautions you should observe are described.

1.1 Vocational Qualifications

The approach to training in industry is changing. Instead of a reliance on knowledge which has typically been assessed by set piece examinations, there is now a focus on competence which is assessed in the workplace. This means that a worker must not only know what he or she should be doing but also has to demonstrate that he or she can do it!

Typically, set piece examinations require the examinee to achieve a 'pass' mark which may be as low as 40%. This accepts the fact that by no means all of the subject matter is known well enough for the examinee to pass the examination. Under the competence system, to qualify, the worker must satisfy the examiner, usually known as the assessor, in all matters. The worker must demonstrate knowledge, ability and communication skills as required for the level of the qualification being assessed. These qualifications are known as vocational qualifications (VQs).

There should be VQs for all employees from the newest and youngest to the most senior. Through training, workers may progress to higher and higher levels. In the developing British system there is a framework of five levels which can be described as:

- **Level 1** – Competence in the performance of a range of varied work activities, most of which may be routine and predictable.
- **Level 2** – Competence in a significant range of varied work activities, performed in a variety of contexts. Some of the activities are complex or non-routine, and there is some individual responsibility or autonomy. Collaboration with others, perhaps through membership of a work group or team, may often be a requirement.
- **Level 3** – Competence in a broad range of varied work activities performed in a wide variety of contexts most of which are complex and non-routine. Often considerable responsibility and autonomy, and control or guidance of others is required.
- **Level 4** – Competence in a broad range of complex, technical or professional work activities performed in a wide variety of contexts and with a substantial degree of personal responsibility and autonomy. Responsibility for others is often present.
- **Level 5** – Competence which involves the application of a

significant range of fundamental principles and complex techniques across a wide and often unpredictable variety of contexts. Substantial personal autonomy and often significant responsibility for the work of others and for the allocation of substantial resources feature strongly, as do accountabilities for analysis and diagnosis, design, planning, execution and evaluation.

It should be possible to categorise all jobs within a company in one of these five levels. To achieve accreditation at any level it is necessary to satisfy the assessor in a defined number of **units** (each of which has a number of **elements**). There are some mandatory (core) units and some optional units, a defined number of which must be selected, for each level. This reflects the fact that workers at a given level may have jobs that are biased towards production or production support and may be more technical or supervisory. In many respects there will be overlaps between levels and the greatest difficulty tends to arise between levels 3 and 4. The flavour of level 3 should be 'optimise, prioritise and improve' and of level 4 'plan, initiate, develop and manage'.

Thus, it can be seen that as the level of competence progresses there is a change from understanding, to seeking ways of improving and helping the business to become more efficient.

For all jobs and tasks there is a need to define What, Why and How? In biscuit making there are technical aspects which include, what are the ingredients, products, processes and machines, why particular ingredients, packaging materials and processes are needed for different products and how machinery is controlled and maintained. Technical knowledge and skills are needed for problem solving, and to ensure good hygiene and safety in the workplace. Communication skills are also needed which include reporting to and supervising others.

Competence cannot be achieved only from a book but reading and thinking are an aid to learning about ingredients, processes and machinery and understanding what variations may occur. Thus, a book can provide essential underpinning knowledge and is a source of reference when something new or unusual happens.

Using this manual will help you to become a competent employee involved in biscuit manufacturing. The underpinning knowledge with respect to making and handling dough that is contained here is relevant particularly to most of the technical aspects of levels 2 and 3, as defined above.

2 Background to the biscuit industry

2.1 What are biscuits?

Biscuits are small baked products made principally from flour, sugar and fat. They typically have a moisture content of less than 4% and when packaged in moistureproof containers have a long shelf life, perhaps six months or more. The appeal to consumers is determined by the appearance and eating qualities. For example, consumers do not like broken biscuits nor ones that have been over or under baked.

Biscuits are made in many shapes and sizes and after baking they may be coated with chocolate, sandwiched with a fat-based filling or have other pleasantly flavoured additions.

2.2 How are biscuits made?

Biscuits are a traditional type of flour confectionery which were, and can still be, made and baked in a domestic kitchen. Now they are made mostly in factories on large production plants. These plants are large and complex and involve considerable mechanical sophistication. Forming, baking and packaging are largely continuous operations but metering ingredients and dough mixing are typically done in batches.

There is a high degree of mechanisation in the biscuit industry but at present there are very few completely automatic production plants. This means that there is a high degree of dependence on the operators to start and control production plants. It is essential that operators are skilled in the tasks they have to do and this involves responsibility for product quality. As part of their training they must know about the ingredients and their roles in making biscuits. They must be aware of the potential ingredient quality variations and the significance of these.

There are basically two types of biscuit dough, hard and soft. The difference is determined by the amount of water required to make a dough which has satisfactory handling quality for making dough pieces for baking.

Hard dough has high water and relatively low fat (and sugar) contents. The dough is tough and extensible (it can be pulled out without immediately breaking), like tight bread dough. The biscuits are either crackers or in a group known as semi-sweet or hard sweet.

Soft doughs contain much less water and relatively high levels of fat and sugar. The dough is short, (breaks when it is pulled out) which means that it exhibits very low extensible character. It may be so soft that it is pourable. The biscuits are of the soft eating types which are often referred to as 'cookies'. There are a great number of biscuit types made from soft doughs and a wide variety of ingredients may be used.

The machinery used to make biscuits is designed to suit the type of dough needed and to develop the structure and shape of the individual biscuits.

Secondary processing, which is done after the biscuit has been baked, and packaging of biscuits are specific to the product concerned. There is normally a limited range of biscuit types that can be made by a given set of plant machinery.

Many biscuit production plants bake at the rate of 1000–2000 kg per hour and higher rates are not unusual. Given this and the sophistication of the production line it is most economical to make only one biscuit type for a whole day or at least an eight hour shift. Start-ups and changeovers are relatively inefficient.

2.3 How a factory is arranged

Typically the factory is long and, for the most part, normally on only one floor. The reason for the length is principally due to the oven.

Tunnel ovens have baking bands that are usually between 800–1400 mm (31–55 in) in width. The length of the oven determines the output capacity of the plant. Ovens have been made up to 150 m in length but 60 m (about 200 feet) is probably the average length.

Ideally, and normally, the ingredients are stored and handled at one end of the factory. Next to the ingredients store is the mixing area and next to that are the continuous production plants. The baking plants feed cooling conveyors, which are often multitiered to save space, and the baked and cooled biscuits are then packed using high speed machines.

In some factories secondary processes are involved after baking. It is also possible that only semi-automatic packaging is used which requires manual feeding of the wrapping machines. In these cases biscuits may be taken from the baking line and placed temporarily in boxes or stored in other ways. These activities are typically labour intensive.

2.4 What your company requires from the factory

Your company exists to make a profit! The means of earning this profit is by making and selling biscuits (and possibly other products).

The products that are made are designed to meet current market needs and to this end they have specifications in terms of pack size, biscuit eating qualities and appearance and ingredient types and quantities. These specifications define limits and it is the task of the production department to ensure that only biscuits which meet the specifications are packed and sold. All substandard product must be disposed of through other routes and will represent a financial loss to the company.

The production cost of a product is a combination of ingredient and packaging material costs, labour (which involves the production time), fuel for the machines and baking, and overheads which include management, maintenance and other support services. A significant cost is the labour associated with the production time. The efficiency of production, which is measured by the quantity of saleable product produced in a specified time, is an important aspect of the product cost.

The duty of the Production Department to the company is to produce a minimum of scrap product and have a minimum of production downtime. Both of these requirements are influenced strongly by the skills and performance of the plant operators.

2.5 Dough mixing areas

The arrangement of dough mixing facilities (ingredient metering and mixers) relative to the main baking lines varies in different factories. It is commonly the case that all the mixers and associated equipment are sited together in one area under the supervision of one foreman. This area may be on a level above the baking plants, and mixers can be arranged to be in line with the respective plants that they feed. The mixed dough can then be discharged directly into hoppers and then into the plants below.

This arrangement assumes that dough is used more or less directly after mixing. In cases where fermentation or some other conditioning period is needed the dough will be discharged into tubs and taken to a place for holding. Alternatively where continuous mixers are used these may be sited much nearer to the baking plants.

2.6 Your contribution when making doughs

The most important contribution you can make to the efficiency of the factory when making doughs is to ensure the following:

- Follow the procedures and timings that have been specified so that doughs are ready as required. The baking plants will not have to stop through a shortage of dough and the dough will be of uniform age throughout the production period. Very fresh or very old doughs will handle and bake differently.
- Understand the appearance and feel of satisfactory dough immediately after mixing is completed. If a change in appearance or consistency occurs you will recognise this and will have a good idea of what may have caused it and what can and cannot be done to rectify the situation.
- If small changes to the formulation or mixing conditions are permitted in the specification and it is within your responsibility to make these changes you must record what has been done, why and when, for future reference by management.
- Keep yourself, utensils and the working area clean and tidy to avoid dirt and unwanted materials getting into the dough. Clean utensils and machinery as specified and especially at the end of a production period or at the change in a formulation.
- Look out for machinery that requires maintenance for reasons of oil leakage, malfunction or breakage and report the problem without delay to your supervisor.
- Avoid straining yourself; do not try to move heavy weights without help or using the appropriate machinery.

In a well managed factory you will probably be required to record various aspects of what you have done. For example, the times that doughs were completed and their temperatures. In this way there will be no misunderstanding about what was done. When and if there are problems later in the process it will be possible to track back and identify exactly where an ingredient or dough originated.

It is not unusual to find faults or to have queries. You must communicate.

It is essential that if you are in doubt you should not hesitate to ask, even if you know that you should know the answer!

3 Hygiene and safety aspects

The regulations relating to food production are continually being tightened with the aim of improving the safety of food products and the safety of people working in food factories.

If you work in a food factory you must be acquainted with the potential dangers and constantly endeavour to prevent the food becoming contaminated with noxious substances. You must also make sure that your actions do not put yourself or your colleagues at risk of injury.

3.1 Safety of food products

Biscuits will be unfit or unpleasant to eat if they are contaminated in the course of their manufacture and packaging. Contaminated means that unwanted material becomes included in or on them. Some forms of contamination may be positively dangerous to the health of those who eat the biscuits.

It is therefore important that the problem of contamination is considered because it is the basis of food hygiene which is the responsibility of all who work with food.

It is not possible to list all the possible hazards to hygiene that may be encountered in a biscuit factory but the following section should help to make you aware of the likely problem areas.

3.2 Sources of contamination

3.2.1 People

- Contamination may come from people via the microorganisms on their hands. Hairs, buttons and pieces of jewellery may fall from their bodies and clothes and articles may fall from pockets.
- The most important requirement for all those who handle, or are likely to handle food, is to observe basic rules of personal hygiene.

Disease is quickly spread if food handlers are negligent about hand washing following visits to toilets. It is very unpleasant to have food contaminated with grease or other dirt from unwashed hands.

- At all food premises good, clean washing facilities must be provided with continuous supplies of hot and cold water, non-scented soap and disposable towels. Cold water with no soap and communal towels are not adequate.
- Hand washing sinks and facilities must be separate from those used to wash equipment.
- **All food handlers must ensure that their hands are washed and clean before handling food and it is particularly important that their hands are washed after each visit to the toilet.**
- Employers must provide clean overalls and hair coverings for all personnel. These should be worn only in the food factory. No personal food, drink containers, loose money, pins, jewellery (other than plain wedding rings), watches, radios, books, newspapers and smoking tackle should be allowed into the production areas. Hair brushing or combing necessitating removal of head gear should be forbidden in production areas. **In this way the possibilities of contamination by loose articles is significantly reduced.**
- Smoking involves the hands becoming contaminated with saliva and the by-products – matches, ash and cigarette ends, are particularly repulsive. **No smoking should ever be allowed in the production areas.**
- Operators who have cuts, abrasions or skin infections, particularly on the hands or arms, should be especially careful. **Bandages or dressings should be of good quality and be, at least partly, brightly coloured and easily detectable should they be lost.**

  In those premises where metal detectors are available for product scanning, it is additionally useful for the bandages to contain metal strips that will be found automatically should a bandage be lost in the product.
- **Food handlers suffering from intestinal complaints** such as diarrhoea or other contagious diseases **should be required to keep away from production areas** until they recover.
- It is frequently necessary for operators to carry certain small articles with them in the course of their duties. **Articles such as pens, pencils, gauges and various tools should not be**

carried in top pockets in case, while bending over, they should fall into the product or machines. Overalls provided with no top pockets remove this possibility!

- Where gloves are needed either of fabric type (as for chocolate handling) or waterproof, they require regular washing and drying both inside and out. **Gloves should not be used by more than one person and they should be replaced when damaged**.

3.2.2 Emptying containers

- When bags or boxes are opened and emptied there is a great potential for contamination.
- Pieces of string or paper removed in the opening process must be placed in rubbish bins and not on the floor.
- Before inverting a bag, box or other type of container, ensure that it has not collected floor or surface dirt that could fall into an unwanted place.
- Dispose of the empty container in a responsible way so that spillage or dust is avoided as much as possible and it is not a danger to other workers.

3.2.3 Small items of equipment

- In most biscuit factories it is necessary to use bowls, beakers or trays to carry and weigh ingredients or dough. These should be made of metal or plastic because glass is particularly dangerous, making splinters or small fragments if broken.
- **Glass containers must never be taken into production areas**. Where ingredients are delivered in glass containers they should be dispensed into non-breakable containers in specially designated rooms away from the production areas.
- Colour coding of containers is better than labels which may fall off. Elastic bands provide a particular hazard due to their tendency to fly off in unexpected directions and become lost.
- All utensils should be stored, full or empty, on special clean stillage so that they are off the floor. This is to ensure that when inverted no floor dirt can fall from them onto product or into a mixer.
- After use all containers should be washed in hot water, with detergent as necessary, and left inverted to dry.

- Cleaning equipment such as cloths, brushes, mops and scrapers should be stored and dried after use on specially provided racks, hooks or rails, off the floor.
- Detergents used for cleaning equipment must be of approved types and stocks must be stored separately away from ingredients or dough containers.
- Office equipment such as elastic bands, paper clips and particularly pins should be forbidden in the factory environment.

3.2.4 Plant machinery

- At the end of each production run all machines should be cleaned immediately so that buildup of dough or other materials does not become hard, or mouldy and an attraction for insects.
- As a basic principle, all food machinery should be mounted off the floor so that the floor can be thoroughly swept or washed at regular intervals.
- Covers for the moving parts of machinery should be properly fixed at all times and kept in good repair.
- All surfaces should be wiped down regularly and washed with warm water and detergent if necessary.
- Fabric conveyors should be checked regularly to watch for frayed edges or seams. If necessary these should be trimmed with a sharp knife or the conveyor replaced.
- If a machine is not to be used for some time it should be covered with a dust sheet.
- Drip trays and other catch containers must be emptied and cleaned regularly, but certainly at the end of each production run.
- Particular care should be taken that mineral lubricating oils and greases do not contaminate food. Leaking motors, gearboxes or bearings should be reported without delay for engineering maintenance.
- Where it is necessary to climb up to high parts of machines or where ladders are needed to get over machines, special walkways with adequate guarding should be provided to prevent floor dirt, carried on footwear, dropping on to dough, products or food surfaces.
- No string should ever be used to attach wires or other articles in production areas and fibrous or loose insulation materials should be covered and fixed securely to prevent disintegration.
- Nowhere in production areas should wood be used. This is

easily splintered and pieces find their way into ingredients or dough.

- As machinery is replaced or becomes obsolete it should be completely removed from the production area and stored (preferably in a reasonably clean condition) in a store remote from the factory. In many factories the machinery graveyard is an ideal home for marauding rats, mice and insects. It is ideal from their point of view because it is dry and undisturbed. Food can be taken there and breeding take place in relative comfort! The convenience of such a home should be denied within the production environment.

3.2.5 Buildings and general factory areas

A major source of contamination is from insects, animals and birds. Also dirt or loose particles falling from overhead areas offer potential hazards.

- Flying insects and birds must be excluded from the factory by using screens over ventilation fans and windows which open.
- Open doorways should have plastic strip or air curtains to prevent entry of insects and birds.
- Doors to the outside should fit closely to the floor so that animals cannot enter at night or other times.
- Rodent control systems should be regularly maintained and any bait must be placed only in specially designed and sited containers which are clearly marked. Damaged bait containers should be disposed of immediately and safely.
- Trunking for wiring and other services should be well sealed to reduce the chance of dust accumulation followed by insect infestation.
- High ledges and roof supports where dust can collect should be of sloping construction and be cleaned regularly.
- Good lighting should be maintained in all production areas and plastic screening, where appropriate, should be used to prevent glass falling on to the product if light bulbs or tubes are broken.

It is a requirement that all food workers are aware of these potentials for contamination and that they report to management, without delay, any aspects that do not seem satisfactory.

3.3 Safety of people

Your employer is required to ensure that the areas in which employees work and the machinery they use is safe. However, if employees are negligent in reporting faults or in cleaning operations, etc. it is possible that an otherwise safe situation can become unsafe. You should therefore be aware of where unsafe situations may arise which could affect you or your colleagues.

3.3.1 Floors

Dirty floors which have become wet or greasy are slippery. Clean up as necessary.

3.3.2 Machine guards

Moving parts especially those where a nip is involved must be guarded to prevent hands or clothing becoming trapped. It is particularly dangerous to run a machine with these guards removed. Experience shows that accidents involving machines occur more often to 'experienced' operatives. They become overconfident and try to overcome problems by running machines with guards removed.

3.3.3 Electrical connections

Most machinery is driven and controlled by electricity. For safety and other reasons the connections and other electronic components are housed in cabinets or under guards. The danger of electricity cannot be seen, so it is very dangerous to remove guards. Faults in electrical apparatus must be reported to management or responsible engineers.

3.3.4 Strain injuries

Back strain is a common injury experienced by factory workers. It is unpleasant for the person who receives it and a potential cost to the employer due to the need for sickness leave.

Back strain derives from physical effort done incorrectly or carelessly. Think when moving and lifting bags, boxes or pieces of machinery. If they are too heavy get help. If the floor is slippery take extra care. Do not try to lift something too high without help. Do not expect a colleague to help you if he or she is not clear what is expected or is not strong enough to do it.

3.3.5 Dust

Dirt is defined as material in the wrong place! Dust soon becomes dirt. It is unpleasant and may be dangerous to breath and it may accumulate and fall into containers bearing food or ingredients. Make sure that your actions cause as little dust as possible and clean up after you.

3.3.6 Building maintenance

When repairs and decoration are being done it is essential that nearby production equipment is covered with sheets so that particles of metal, glass, masonry and paint do not fall into places where they could later be included with dough or product. A magnet provides a useful means of collecting most metal particles.

4 Problem solving

4.1 Introduction

Biscuits are usually baked in vast numbers in continuously running ovens. It is necessary to maintain a constant vigilance to ensure that the biscuits are made with minimum variation. If this is not done, or is done badly, biscuits may be made that are not suitable for consumption or that will not fit into the packaging. From time to time it is necessary to make adjustments and alterations to the process or dough formula to maintain the biscuits within specification. It is the duty of plant operators to make these adjustments. The efficiency of a factory is largely measured by the skill of operators in making these adjustments correctly, accurately and speedily.

In most cases problems are not caused by one factor alone. They are the result of interactions between several factors including ingredients, machine settings and other processing conditions. This means that check lists for both identifying and solving problems are often not simple. Ideally it would be good to have an interactive problem-solving computer program that would interrogate the operator(s) and offer a set of instructions that would lead to the removal of the problem. Not all factories have computers available on the factory floor. Where computers are not available, it is suggested that **reference charts and lists** are kept as a back up to experience and memory of what happened previously. Having observed a problem, the chart or list can be used as a guide to further action. However, understanding why a fault or problem may have arisen is better assessed with some background knowledge. Reference lists to problem solving may be found in each section where a process is described and a reference can be found on each problem solving chart to the section in this manual that provides more information about the parts of the processes involved.

4.2 Process audit record

Much difficulty is encountered at plant start up. A late start is inefficient and can be costly to the factory. There is therefore a need to set the plant so that the start up is as smooth as possible. Drawing on previous experience is clearly the best approach and to this end a process audit record (see Fig. 1) should be available for each plant for each biscuit that is made. It will give details of biscuit recipes, mixing conditions and process measurements that have to be taken. From time to time the validity of the information on the process audit record will be checked and reviewed because processes can change as a result of time of year and quality of ingredients, etc.

4.3 Control philosophy

When faults and difficulties occur it is necessary is to determine their cause and to act as quickly as possible to cure them. The aim is to enable quick fixes, not detailed research programmes that will take a long time and lead to significant changes in processing methods. It is assumed that production of the products in question has proceeded well for long periods but problems have arisen, which if not attended to, will cause inefficiency or waste.

First, the nature of the fault or problem must be identified. If a pattern in the occurrence of the fault can be found, its cause, and thence its remedy, is more likely to be found. The nature of the patterns of faults and problems will be discussed as appropriate.

If the control philosophy is wrong, the order of making adjustments will not solve the problems that are encountered satisfactorily. Although this manual relates to dough preparation and handling it will be appreciated that this part of the biscuit making process cannot be viewed in isolation. Therefore, an overall statement of control philosophy follows.

Control implies adherence to product specifications and production targets. The data on the process control chart will show details of the plant and process when, on a single occasion, the whole plant, not just the dough piece forming area, was running well. It will therefore be a useful reference point if difficulties are encountered.

The following principles should be observed:

- Make measurements, but only where and when required.

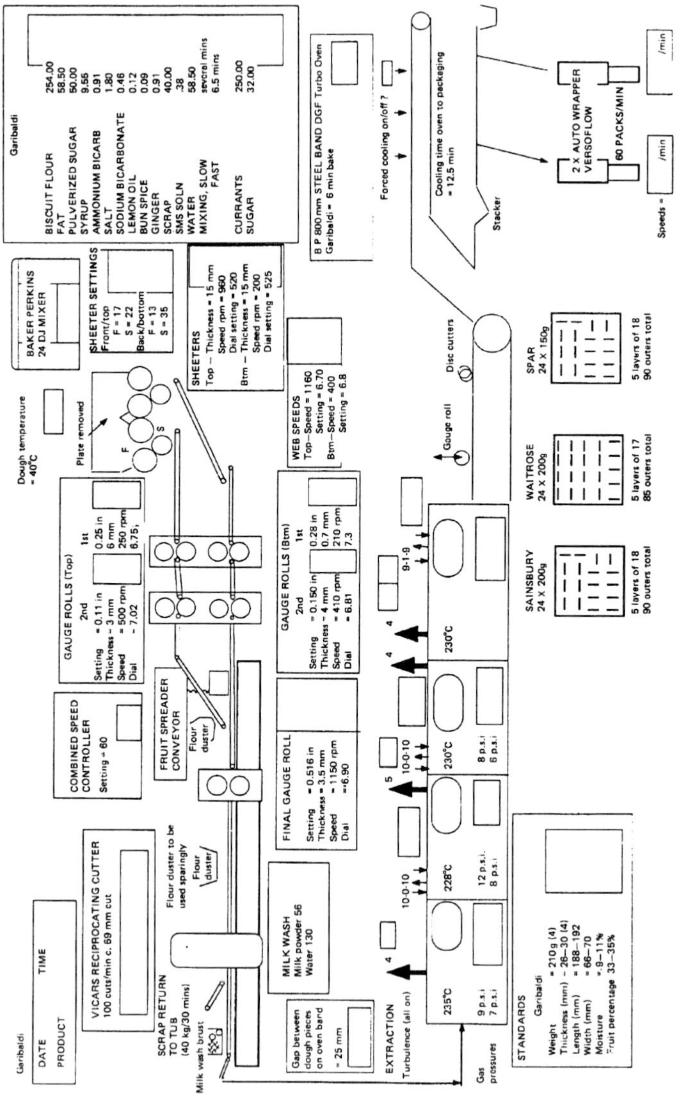

1 Process audit record for production of Garibaldi biscuits

- Relate sampling and measurements to subsequent processing requirements.
- Consider the significance of the results of measurements against those that can reasonably be achieved with the instruments being used. For example, if the reproducibility of a test is poor then small differences in measurements may not mean that changes are occurring.
- Relate measurements to targets which have specified limits.
- Try to understand what affects the parameter being measured and therefore what type of action can be taken when necessary.
- Concentrate process control surveillance on baked biscuits at the oven exit.
- Biscuit weight is the primary parameter. Changes in weight will probably result in changes in other measurable characteristics like size and colour so concentrate on controlling weight before other characteristics.
- Concentrate process control surveillance at those places where significant variation is known to occur or can be expected.
- Record all measurements against time preferably on a chart which allows easy view of changes and trends.
- Record **all** changes in formulations including water level changes.
- Record the temperatures of doughs at the completion of a mix and also after periods of fermentation or enzyme treatment.
- Review product specifications and process control procedures regularly in the light of process control records.

The measuring tools applicable to dough preparation and handling are:

- weighers with appropriate scales for metering ingredients and scrap dough,
- a thermometer for checking ingredient and dough temperatures,
- a clock telling the time of day,
- timers for durations of mixes, etc.,
- specialist instruments for assessing dough consistency.

5 Types of biscuit dough

5.1 Introduction

The basic ingredient of nearly all biscuit doughs is flour. To this is added sugar, fat and possibly a wide variety of other ingredients. The dough is bound together with water. As more fat is included in the recipe less water is required.

After mixing, the dough is divided up into small pieces of particular shape and size. These pieces are baked into biscuits. Whether the dough is processed into pieces by machine or by hand the consistency is very important.

The consistency, the feel, of dough is determined principally by the amount of water present and also by the temperature of the dough. Wetter and warmer doughs are softer. The mixing of water with flour involves a number of changes but basically the hydration of the components of the flour is a relatively slow process. This means that some change, which affects the dough consistency, may continue to occur after the mixing has been completed. Thus, when considering the making of a biscuit dough one must think not only about what happens in the mixer but also in the period afterwards.

There are basically two types of biscuit dough, hard and soft. The difference is determined by the amount of water required to make the dough. When there is a high water content (and therefore a relatively low level of fat) blending with the flour involves the formation of gluten from the hydrated proteins. The formation of gluten requires mechanical work, kneading, and thus we say that hard doughs are 'developed' doughs, referring to the mechanical action.

5.2 Developed doughs

Hard, or developed doughs, are used to make crackers and semi-sweet biscuit types. The fat and sugar levels are relatively low. In all cases sustained mixing action is required, to develop the gluten. In the case of cream crackers and soda crackers the gluten is further modified after mixing during a period of fermentation with yeast. However, for semi-sweet biscuit dough all the gluten development takes place in the mixer and the mixing time is relatively long.

We will consider the cracker dough that is to be fermented first. The yeast cells are a living fungus and the speed of their growth and action to produce carbon dioxide gas is determined largely by the temperature. It is therefore important, in the first instance, that the dough is mixed to the desired final temperature (usually a value between 26°C and 30°C). This can be achieved both by the length of the mixing time (during mixing a dough warms up through friction) and by the temperature of the ingredients (particularly the water) used.

In some cases cracker doughs are remixed after a period of fermentation and at this time more flour and water is added.

Other cracker doughs involve the use of an enzyme, known as proteinase, which chemically modifies the gluten during a standing period after mixing. The reaction is not too dissimilar to fermentation by yeast, as living cells all use enzymes to effect metabolism, however, in this case no gas is produced. The temperature of the dough is again important as the speed of chemical reaction is very sensitive to temperature. The temperature of proteinase doughs is usually about 35°C.

Typically all types of cracker doughs are mixed in only one stage of mixing. All the ingredients are loaded into the mixer and the mixing proceeds. This is known as an 'all in' mix procedure.

Semi-sweet doughs are different from cracker doughs. They have appreciably more sugar and fat in the formulation and they are not fermented after mixing. A considerable amount of work is needed before the gluten is in an optimum condition. Much research has been done to determine conditions that produce a well mixed dough. The conclusions are that doughs should be mixed to a specific **final temperature**, and provided that the mixing time is above a low minimum, the time to produce the mix is not critical.

In many cases the gluten quality is additionally modified during mixing by using a small quantity of the compound sodium metabisulphite. If this is used, the optimum final temperature of the dough is about 41°C. If this chemical treatment is not used, the final dough temperature should be higher at about 45°C.

The use of sodium metabisulphite, SMS, has a dramatic effect on the dough quality. Typically the level of usage is about 0.029 units to

100 units of flour. Compared with doughs with no SMS the amount of water needed is reduced by 10% or more. The SMS reduces the elasticity of the gluten so the dough shrinks less after cutting. It is therefore easier to control the shape of the biscuits baked from the dough. By adjusting the amount of SMS used some variations in the quality of the flour can be compensated for.

Typically semi-sweet doughs are mixed in only one stage of mixing. All the ingredients are loaded into the mixer and the mixing proceeds.

5.3 Short doughs and batters

As a group, short doughs vary greatly in their formulation. In no case is the hydrated flour protein developed into cohesive gluten. A dough that pulls apart easily and has minimal elasticity and extensibility is said to be 'short'. The formation of gluten is hindered or very limited because the level of fat is high and the level of water low.

The challenge in making a short dough is to disperse the fat and water evenly through the flour and to prevent hydrated proteins forming gluten. Normally this is achieved by forming an emulsion of the fat and the water first and then mixing this quickly with the flour. In the first stage of this procedure the sugar and most of the other ingredients are also added. This allows the sugar to dissolve in the water and all ingredients to become well dispersed. Often the level of sugar is too high for it to dissolve completely in the available water and undissolved crystals of sugar will be an important part of the texture of the baked biscuit.

Typically the first stage will take several minutes of mixing but the time is not critical. There is little or no development of heat as the mixture is very soft. The second stage, after addition of the flour, will be very short, perhaps as little as one minute, depending upon the efficiency of the mixing action. The time is critical in preventing the development of gluten and thus toughening of the dough.

It will be appreciated that during the second stage of mixing there is insufficient time for the water to become totally absorbed into the flour. Hydration continues passively after the mixing has finished. This means that the dough is much softer at the end of mixing than it is after 30 min standing. Allowance must be made for this change in the dough handling and it is referred to again in Section 9.

The most critical aspects of mixing short doughs are first to form a good emulsion of water and fat and in the second stage of the mixing sequence, to limit the formation of gluten. It is possible to adjust the amount of sugar that is dissolved by adding at least some of it with the flour rather than at the first stage of mixing. The proportion of sugar that is dissolved in the dough has an effect on the texture and shape of the baked biscuit.

Some short doughs are very soft and even pourable.

The temperature of short doughs is determined largely by the temperature of the flour, sugar and fat. There is little possibility of controlling or adjusting the temperatures of the flour or the sugar and the fat is normally used soft at about 26°C. A small measure of control of dough temperature can be achieved by using cold or warm water and by holding the jacket of the mixer at a desired temperature. The ideal temperature for a completed short dough is between 18–22°C.

Some 'biscuits' are made from very fluid batters, normally these are based on egg and are really variants of cake formulas. These batters are mixed with high sheer mixers in order to incorporate air bubbles which form the basis of the baked structure.

6 Types of dough mixer

6.1 What is required of a mixer?

The term 'mixing' covers a number of distinct operations. It includes:

1 Blending together ingredients to form a uniform mass.
2 Dispersing a solid in a liquid, or liquid in a liquid.
3 Solution of a solid in a liquid.
4 Kneading of the mass to impart development of gluten from flour proteins which have been hydrated at an earlier stage of mixing.
5 Buildup of temperature as a result of work imparted.
6 Aeration of a mass to give a lower density.

One or more of these actions is involved in the formation of different types of dough. The problem comes because there is no one design of mixer that performs all of these functions with maximum efficiency. For example, a mixer that is good at kneading dough will be poor at forming a fat/water emulsion.

Different emphasis is required for certain 'mixing' functions depending on the formulation and type of dough. As biscuit mixers are normally required to mix more than one formulation it is inevitable that compromises in efficiency or effectiveness must be accepted.

Most mixers perform differently depending upon how full they are.

Normally there is little indication of how mixing is proceeding. Most mixers are totally enclosed preventing observation of the progress of the mix. There is no external indication of how uniform the blend is or how good is the emulsification. In many cases there is no reliable or continuous indication of dough temperature. There are normally timers with which to set the length of each stage of the mixing but otherwise the control possibilities are rather limited!

As a result of this situation there is considerable reliance on observations of the operator to ensure that doughs are produced satisfactorily.

The following procedures are recommended:

- Keep the size of each mix as constant as possible. This will involve not only using a standard quantity of flour and other ingredients as determined by the formulation but also using a constant quantity of rework material, such as old dough, if necessary.
- Mix to preset times for short doughs but watch for lack of uniformity within the dough if the second stage is very short.
- For developed doughs mix to given final dough temperatures. This will be related to time but at the end of the mix check the temperature and if necessary mix for longer.
- If the water jacket of the mixer is used, set the temperature of the circulating water to either the final temperature of the dough, so that a finished mix can be retained in the mixer until needed or, if it is used to cool or warm the ingredients during mixing, discharge the dough immediately the mixing is finished so that dough in contact with the mixer bowl does not become warmer or cooler than the rest of the dough.

6.2 Biscuit dough mixers

There is a surprisingly large number of manufacturers of dough mixing machines. The mixers vary in size, sophistication of control, power and overall weight. However, they can all be classified into a few types and the aim here is to indicate these giving some of the major advantages and disadvantages.

6.2.1 Batch mixers

These are by far the most common type.

6.2.1.1 Vertical, detachable bowl, mixers

The beaters are mounted vertically and either they and their drive mechanisms are lowered into a bowl or the bowl is raised to locate with the beaters and a lid. The beater shafts may rotate in fixed positions, in which case there are usually two or three beaters which intermesh with each other, or there is a single shaft which rotates vertically and itself is driven in a circular, planetary, manner. This action allows a single beater to reach all the dough in the bowl without merely moving it in a circular motion.

It is sometimes possible to fit exchange beaters of different shape and action and to drive them at different speeds. This allows a gentle rolling and cutting action at one extreme and a vigorous whisking action at the other.

In small vertical mixers the beater speeds can be fast but in larger ones, mixing up to one tonne of dough at a time, the beaters revolve slowly, probably not more than 20 rev per min. Motor powers of up to 70 hp may be used. Figure 2 shows a typical vertical spindle mixer.

The **advantages** of vertical mixers are:

- The bowls can be filled with ingredients at various places, away from the mixer, so that loading and discharge is not a critical feature of the mixing time cycle.
- Doughs which need to be rested, fermented or remixed can be left in a tub without additional handling.
- Different mixing actions can be achieved either by using more than one mixer or by changing the beaters on a single mixer.
- In many cases the mixing action and the state of the dough can be visually monitored.
- It is easy to charge the bowl with awkward ingredients such as scrap dough or biscuit dust manually.

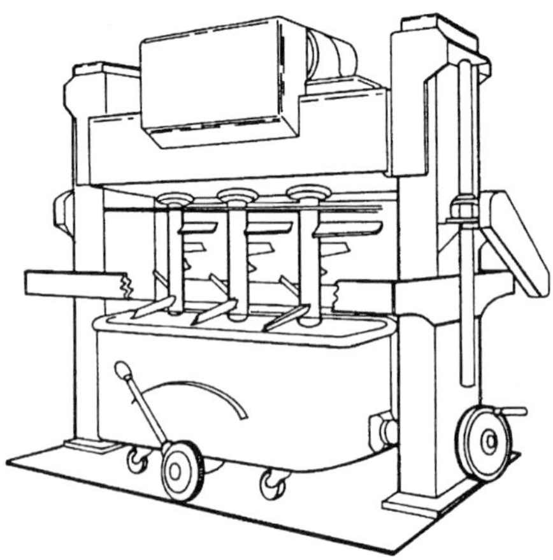

2 A typical vertical spindle biscuit dough mixer.

- Bowls containing dough can be readily moved to different locations for tipping or storage.
- High speed mixing with a whisk allows thin batters to be mixed and aerated.

The **disadvantages** of vertical mixers are:

- The mixing action is sometimes not uniform between the bottom and top of the bowl resulting in more or less development in some parts of the dough.
- The water will always run to the bottom of the mixer before mixing starts.
- It is difficult to maintain good temperature control of the bowls because jackets containing circulated water must be connected and disconnected.
- Dough temperature sensors are rarely provided as these must also be connected and disconnected.
- The bowls are heavy and not very manoeuvrable, sometimes requiring mobile power units, such as forklift trucks, to move them.
- Mixing times are often long because the beaters rotate slowly. It may take as much as 45 min to mix a developed dough.
- The slow mixing action may be an advantage for short doughs as it is less easy to overmix by accident.

6.2.1.2 Horizontal high speed mixers

There is rather less variety in the design of horizontal high speed mixers. There are some where the bowl is fixed and a door at the side or in the bottom opens to allow discharge of the dough, but more usually the bowl rotates on a horizontal axis around the beaters to allow discharge of the dough. The beaters are driven horizontally within the bowl and are fixed to one or two shafts. Where only one shaft is employed the beaters are usually inclined so as to throw the dough not only upwards but also to one side and then to the other during rotation. The blades may pass close to the bowl surface or at some distance. The former type ensures that material lying in the bottom of the bowl is moved but the latter is better for kneading, rolling and stretching the dough. The action whereby the dough is cut and sheared depends on the exact shape and speed of the blades, but sometimes a stator fixed to the bowl provides an additional means for cutting the dough to achieve good homogeneity (see Fig. 3).

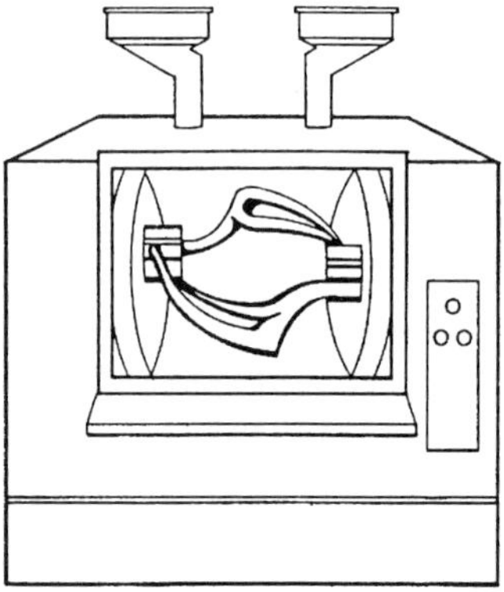

(a)

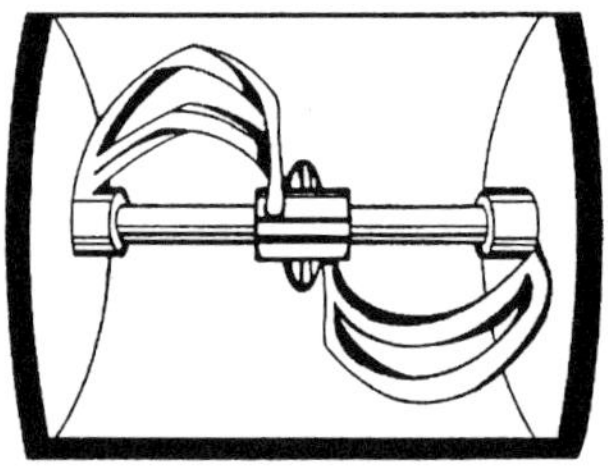

(b)

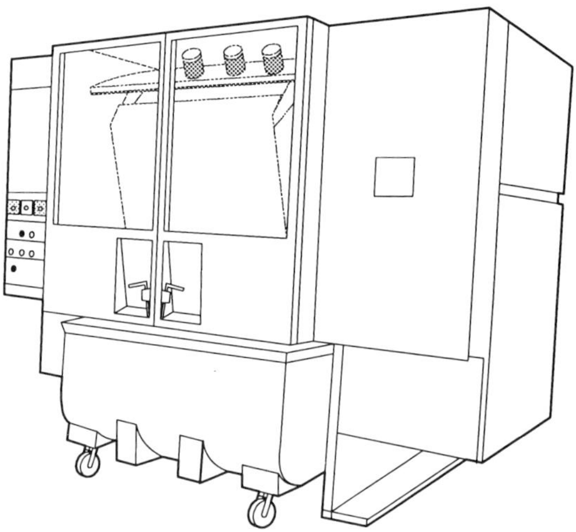

(c)

(c)

3 Typical single blade horizontal high speed mixers.

Where two shafts are employed, the bottom of the bowl is W-shaped and the shafts rotate in opposite directions driving the dough towards the centre and downwards across the centre of the bowl or in the other direction. These mixers are usually known as Z-blade mixers (see Fig. 4).

The beater speeds can be as high as 60 rev per min which means that doughs can be mixed in relatively short times, possibly in 12–15 min for developed doughs. Such mixing actions for loads of up to one tonne require very powerful motors; up to 180 hp are not unusual.

The **advantages** of horizontal high speed mixers are:

- These are very powerful mixers as the shafts have bearings at each end and they are, therefore, able to develop tough doughs more rapidly than most of the vertical types of mixer.
- Provided that the discharge is efficient, it is possible to locate the mixer directly over a sheeter hopper and this obviates the need for handling the dough via a tub.
- Tubs can be used should it be necessary to take the dough to another location.
- There is good and accurate control of the temperature of the

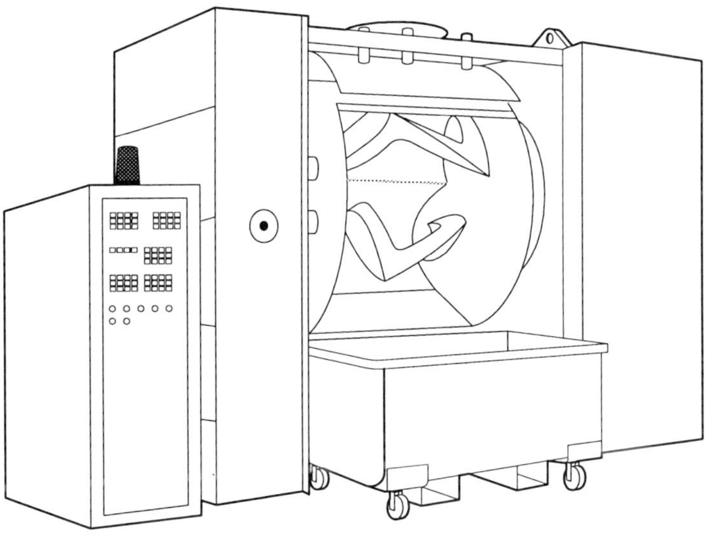

4 Typical Z-blade twin bladed horizontal high speed mixer.

mixing bowl from a jacket with constantly circulating water or refrigerant. However heat exchange possibilities are limited.
- Ingredients can be added through the roof of the mixer while the beaters are moving, that is, after the mixing has started.

The **disadvantages** of horizontal high speed mixers are:

- Charging with ingredients usually constitutes a significant time period in the mixing cycle time and all ingredient feeds must be located over the mixer or be done by hand.
- The beaters tend to throw material up to the roof of the mixer which can result in blind spots where ingredients hang up.
- The lid totally encloses the mixer so that the progress of the mix cannot be readily observed.
- Cleaning is a major operation and because there are some unswept areas of the bowl a 'scrape down' part way through a mix may be necessary.
- It is very inconvenient to remix a dough, since charging with dough from a tub is a particularly difficult operation.
- The beater shape is usually a compromise to allow blending, dispersion and kneading so that each action may not be ideal.
- Where a central shaft crosses the mixer there is often a severe impediment to efficient and rapid dough discharge and this shaft may prevent the free movement of the dough resulting in barrelling, that is, the dough is rotating but is stuck to the beater assembly without achieving much mixing action.
- The mixer itself is heavy and the beating action can give much vibration. This makes important structural demands on the floor upon which the mixer is located, particularly if it is not on the ground floor of the building.
- As the size of these mixers increases the efficiency in terms of kneading action tends to decline relative to the heat buildup due to friction between the dough and bowl surface.
- Large motors require heavy electrical supplies and starting under load draws very heavy currents.

6.2.1.3 Size of batch mixers

There tends to be some confusion on the capacities of various batch mixers. They may be described in terms of bowl volume or in dough weights. It is important to establish, either by experiment or by consultation with the mixer supplier, the maximum (and minimum) dough weights that may be mixed efficiently. Limitations may be related to motor power or to the areas swept by the beaters. It is possible that the dough quantities will be different for different types of dough.

The capacities may be based on 'sacks' of flour (280 lb or about 125 kg) but this is not very helpful where large amounts of sugar, fat and other ingredients are included in the recipes. Alternatively, capacities may be based on volume such as 100, 200 or 500 litres. As a rough guide, these volumes relate to recipes containing 32, 64 and 192 kg of flour. A common size for a batch mixer is 500 kg of dough but others up to about 1000 kg are available.

6.2.2 Continuous mixers

These are generally of a rotor-within-a-barrel variety. By arranging different arms and stators along the length it is possible to alter the mixing actions within the range of blending, dispersing, aerating and kneading. Multisection water jackets allow excellent heat exchange and temperature control and by adjusting the barrel length, dough retention and mixing times can be suited to the optimum. The overall capacity of the mixer is usually flexible by adjusting the rotor and ingredient feed speeds. It is possible to feed all the ingredients at the start of the mixer or to have successive ports along the barrel so that different additions can be made after appropriate intervals.

The **advantages** of continuous mixers are:

- Without doubt, continuous mixers provide the best and neatest facilities for making doughs or batters in an optimum way.
- Their output can be precisely matched to the rest of the production plant so that all dough is of uniform age.
- When they are running, minimum supervision is required.

The **disadvantages** of continuous mixers are:

- Starting and stopping is not easy and presents a problem should the rest of the plant stop for any reason.
- Setting up continuous mixers is difficult and presupposes knowledge of optimum mixing conditions and sequences. For this reason, use for a range of different recipes may be difficult as each may not require the same conditions.
- Metering of all the ingredients must be continuous and this equipment and its maintenance can be extremely costly.

- It is not easy to meter scrap dough in a uniform and constant manner. To reduce the number of ingredient feeds it is best to consider the preparation of premixes both of solids and liquids and this may be a major capital and supervisory expense.

Continuous mixers are best considered for single purpose plants.

6.2.3 Automatic small batch mixers

As a compromise between batch mixers and continuous mixers, automatic small batch mixers have been introduced. These have ingredient feed systems that are totally electronically controlled and the mixing cycle is triggered to start by reference to the level of dough in a hopper directly below the mixer.

The cycle starts by calls for all of the ingredients. After the mixing period the bowl inverts completely and discharges into a hopper below. When the dough has been used and the level drops to a preset level the cycle begins again. There is no human involvement, except when something goes wrong!

The size of the mixer and the quantity of dough that is mixed each time is chosen to be as small as possible so that many small batches are produced successively. In this way the age of the dough as it is processed through the baking plant is very nearly uniform.

It is normal to design in a sensing system that can be used to compare conditions in the mixer, for example temperature and or power usage related to time, against a norm. If conditions do not correspond the dough is not dropped into the hopper, an alarm is sounded and an operator must check that the dough is satisfactory to discharge normally.

The **advantages** of automatic mixers are:

- The mixer uses actions with which one is already familiar on standard batch mixers.
- The sequence may be stopped at any point to allow visual inspection of the material in the mixer.
- It is easy to reprogramme the system with a different formulation or with different mixing times and speeds.
- Being intermittent in action, stops on the baking plant do not require special attention to the mixing operations.
- If a fault arises dough of substandard quality can be easily removed from the system provided it is detected before dropping into the hopper.

There are few **disadvantages** compared with continuous mixers:

- The machinery is larger but not necessarily more expensive.
- There is no provision for hand feeding ingredients. This means that there may be some limitations in formulation where special, difficult to handle materials, are used.

7 Mixing the dough

7.1 Production control considerations

To run efficiently, baking plants require a continuous supply of dough which is of uniform quality. Stops in production cause many problems and are therefore costly. Stops may occur if the dough supply runs out, even temporarily, or the consistency of the dough is unsuitable for the dough piece forming machines. It is therefore a key requirement for the mixing operator to provide dough as required, neither too soon nor too late, and to ensure that the dough consistency is similar and satisfactory from one batch to the next.

For each type of product, mixing instructions and dough specifications will have been determined and will be provided for the mixing operator. From time to time changes in procedure will be required. These may be as a result of plant stops, variations in ingredient properties that necessitate small adjustments to the formulation, or mixing times, etc.

It is easy for a small problem to become a large one if records are not made and actions communicated as required (see Fig. 5). If the mixer operator works in isolation from his or her colleagues on the baking plant that operator is not fulfilling his or her duty with respect to product quality.

As an operator becomes more familiar with the process and generally more competent he or she may be tempted to 'help the efficiency' by making changes that he or she is confident will be right but for which no authority has been given. Problems can arise because this action may not be in accord with what others are doing and the combined effect may result in unexpected difficulties. Therefore any suggestions of change that fall outside given authority should be agreed before they are initiated. Changes should be kept to minimum and should only be made based on reliable measurements and against preset limits. All changes must be recorded and communicated forward down the plant so that no one has a surprise!

MIXING RECORD SHEET

Date _ _ _ _ Formulation name _ _ _ _ _ _ _ _ _ Mixer no._ _ _ _ _ _Mixing operator _ _ _ _ _ _ _ _ _ _ _ _

Date of previous production_ _ _ _ _ Water level at end of that production _ _ _ Litres

Standard dough temperature_ _ _ _ °C Dough temperature at end of previous production _ _ °C

Weight of scrap available from previous production _ _ _ _ _ kg

					Authorisations for recipe changes (Enter only if a change has to be made)			
Mix no.	Time mix completed	Dough temp. °C	Weight scrap dough used, kg		Water level, ltrs	Other ingredients	New weight	Signature
1								
2								
3								
4								
5								
6								
7								
8								
9								
10								
11								
12								
13								
14								
15								
16								
17								
18								
19								
20								
21								

Weight of scrap dough remaining for use in next production _ _ _ _ _ kg

5 Typical mixing operator's record sheet.

Changes to **dough water levels** are not uncommon and are related to ambient temperature conditions and changes in flour properties throughout the year. There would not expect to be many changes during one day or shift. If there are many changes, there could be a malfunction in the ingredients metering system which it is important to know about and correct if possible. From the records of problems and changes both management, technologists and colleagues on the next shift will know what the current conditions are and where to be vigilant.

Incorporation of **scrap dough** should be made with care and understanding. The inclusion of scrap dough never improves the dough quality. Scrap dough arises either from the end of a previous production run or from faulty mixings that cannot be passed for production of biscuits. In a few cases the scrap may be cutter scrap, that is, it has been returned to the mixing department from the forming machines. Generally, scrap dough is harder, cooler and drier than fresh new dough and it may be necessary to increase the water level used to mix the dough with which it is incorporated.

Scrap dough is sometimes included in mixes because it would be costly not to use it. By no means all scrap dough is in a suitable condition for reincorporation.

Always use scrap dough in specified quantities and in the specified way. This usually means that only a small quantity can be a incorporated in a mixing and while scrap is being used up the **same** quantity should be added to each mixing. The quantity of scrap dough taken should be **weighed** each time to ensure that the quantities are similar for each mixing. Variations in the quantities of scrap used will result in variations in biscuit quality.

Normally only scrap dough of the same recipe as the fresh dough is incorporated. Always put scrap dough into the mixer as the first 'ingredient' to permit maximum dispersion to occur during mixing. This is the case for both developed and short doughs.

Always keep a record of how much (and the type of) scrap that is used in each dough and be sure to communicate to the dough piece forming department when scrap is being used and when no scrap is being incorporated.

As a major contribution to factory efficiency control it is also important that breakdowns and all difficulties that have affected dough production are recorded accurately and their duration noted.

7.2 Dough consistency

The term 'consistency' covers all those aspects of a dough that can be felt in a dough such as resistance to deformation and stickiness. Characteristics such as softness, plasticity, elasticity, stickiness and pourability can all be assessed when a lump of dough is squeezed or pulled. In common with most other materials, as the temperature of dough is increased it becomes softer. Thus, temperature is another character that can be felt and measured in a dough and which is used in an assessment of consistency.

Biscuit doughs are complex being made up of a liquid phase that will be of fat and water and a solid phase that includes starch, protein, sugar and often many other materials. Some of the fat may be solid and some of the solids may be dissolved in the water. Furthermore, doughs change on standing. In some cases the water is absorbed slowly by an ingredient after the end of the mixing period (oat flakes are a good example of this). This results in a tightening or hardening of the consistency. In other cases, the elasticity of the hydrated and mechanically developed flour protein, then known as gluten, becomes less extensible on standing and this results in a significant change in the feel and behaviour of the dough. In general, dough that has just been worked or moved either by the mixer or by handling has a softer consistency than one which has been standing. This property is known as thixotropy (a well known example of thixotropy is in 'non drip' paints which are thick as they are taken from the container and become more liquid as they are worked on a surface with a paint brush).

It is very difficult to measure dough consistency and to give it a value. This is principally because of the effect of handling the dough before the measurement is made. Just the act of putting a sample of dough into an instrument can have a significant effect unless very strict procedures are followed. Furthermore, most instruments that can be used to measure consistency are too delicate to use in the rugged environment of the biscuit mixing department. For this reason it is almost impossible to give a specification for the consistency of any particular dough and considerable reliance is made on the ability and experience of the mixer operator to 'know' his dough and to be able to detect differences and spot changes and faults.

However, the consistency of the dough is of great importance to the smooth running of dough piece forming machines. These machines press and roll the dough continuously so that changes in the consistency and stickiness have significant effects on performance. The smooth running of the forming machinery depends on a uniform quality of dough.

7.3 Troubleshooting

7.3.1 Mixer sequencing programme

Where bulk handling of any of the ingredients is involved it is usual to have the sequence of loading the mixer and mixing the dough controlled automatically.

Elements of the sequence and its control by the mixer operator will include:

- setting the calls for the various bulk handled materials to be metered and eventually dropped into the mixer bowl
- setting the timers for the duration of mixes at selected speeds
- starting the sequence.

As the sequence proceeds there may be:

- prompt lights or sounds to indicate that manual additions of ingredients are required. A resume button will allow the sequence to continue after the additions have been made.
- an audible alarm to indicate a malfunction
- an indication that the sequence has finished and the dough is ready for discharge
- a switch to allow individual operations to be called and initiated manually instead of through the automatic sequence.

7.3.1.1 Action in the event of stoppage of the mixer sequencing programme

Progress through a sequence does not always proceed as it should and it is therefore necessary that the operator should know what action to take to complete the dough satisfactorily should a fault occur.

Details of the sequence depend on the way in which the bulk materials are metered into the mixer and also by the type of dough being prepared. The following comments and suggestions for dealing with problems are therefore of a general nature only. There may be several other types of problem associated with particular installations.

If the operator notices that the mixing sequence has stopped and apparently nothing is happening he or she should make some checks before taking any action.

- The most usual cause of a stoppage is that the sequence is demanding some **operator intervention** such as the manual loading of a small ingredient or to allow inspection of the dough. (Often a mixing sequence is programmed to proceed to a particular point and then to wait for an instruction to proceed. This is to ensure that the mixing is not completed before it is wanted.) Such stoppages should be indicated by prompt lights but it is important to be sure at what stage the sequence has reached so that the correct action is taken. After the required action has been taken press the resume button.
- Another common cause of a stoppage or delay is that the **bulk handling system is busy** and the call for an ingredient is in a queue. Check which ingredient is required and try to establish whether there is a fault or whether the ingredient is being delivered to another place. Other problems associated with ingredient metering are described in Section 8 (see Fig. 6).

7.3.2 Action in the event of dough quality problems

7.3.2.1 Developed doughs

The most likely problem to be encountered is that the dough consistency is unsatisfactory at the end of the mixing. In view of what has been said about dough consistency it is difficult to define what is 'unsatisfactory'. An appreciation must be learnt of what type of dough and what size of variation in the feel of a dough is acceptable for the next stage of the process, usually, dough piece forming.

If, at the end of the mix, the dough appears unsatisfactory it is essential to try to understand why the situation has occurred. The problem is usually associated with poor metering of the ingredients, for example, not enough flour or not enough water or fat. It is usually too late to know the real cause of the error unless there is a data logging system linked to the ingredient metering equipment. If a faulty dough has been made it is necessary to keep a very close watch on the metering and the whole mixing sequence during the next dough mixing cycle. Remember that it is not unusual for a shortage of flour in one mix to be followed by an extra quantity in the next! This is because some of the flour that should have been

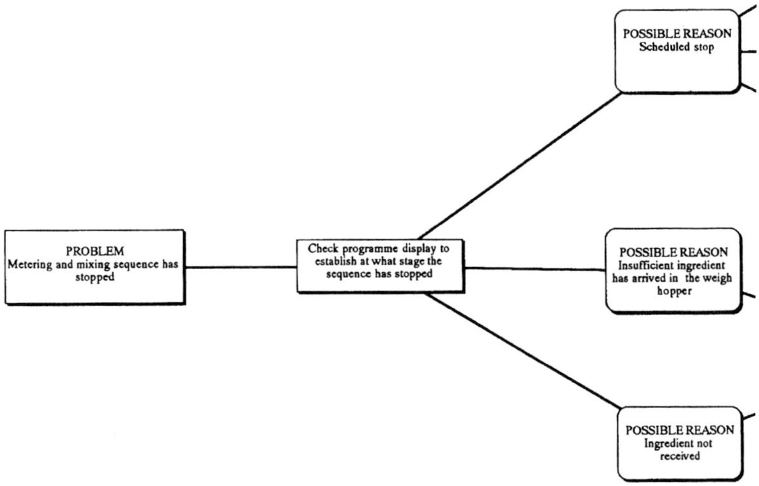

6 Problem solving – actions in the event of stoppage of the mixer sequence programme.

delivered to the mixer became held up and was released with the next delivery.

- First check that the dough temperature is within defined limits. If it is low and the dough is too tight continue mixing until the correct temperature is reached and check the consistency again. With developed doughs there is the possibility to correct the consistency before discharging the dough from the mixer. If the dough is at the desired temperature and is too tight, addition of some water followed by about two minutes of extra mixing is permissible. However, if the dough is too soft it is possible to add some more flour and to remix for three or four minutes. Care must be taken that the flour is completely dispersed by the end of the extra mixing period. The quantities of water or flour that should be added in this way are a matter of experience but the suggested amounts are up to 10% more water or up to 5% more flour (see Fig. 7 and 8).

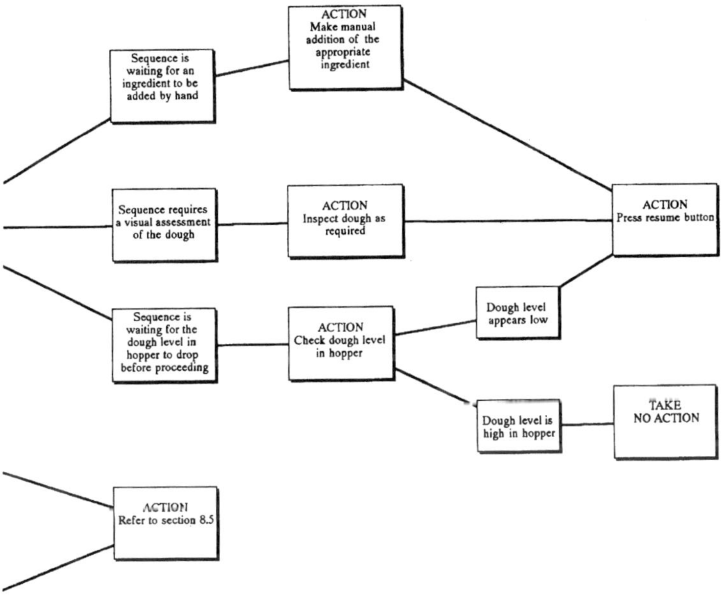

- Another aspect of control that is possible with developed doughs, as has been mentioned in Section 5.2, is chemical adjustment of the gluten quality. If SMS is used for this it is possible to increase the amount of SMS very late in the mixing. The reason for changing the SMS treatment level is prompted not by a dough test or a personal assessment of the dough but by feedback from the baking plant.

7.3.2.2 Short doughs

Unlike developed doughs it is not satisfactory to make a late adjustment to the water level in a short dough to correct the consistency. This is because the last stage of mixing is very critical and attempts to add water at a late stage will change the nature of the dough in addition to changing its consistency.

A short dough becomes tighter and less sticky on standing. These changes occur principally within the first 30 min after the completion of mixing and are the reason why it is recommended that short doughs are stood for a specified time before use.

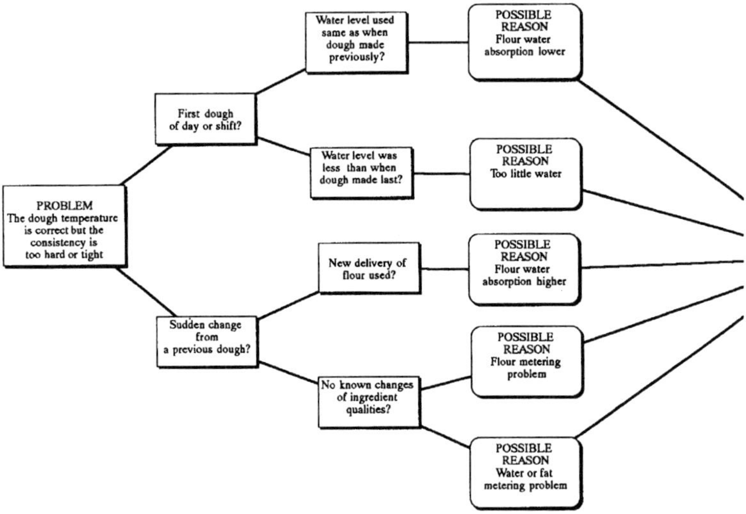

7 Problem solving – developed dough that is too hard or too tight at the end of the mixing programme.

Generally, if a short dough is **tighter** or **drier** than usual and is regarded as being unsatisfactory for subsequent processing, the consistency cannot be satisfactorily adjusted. The dough should not be passed for processing. If the problem is borderline it may be possible to use it without a standing period in the hope that satisfactory biscuits can be made from most of it. Perhaps half of the dough mix could be used in this way (see Fig. 9).

Dough that is **too soft** may become usable if it is given extra standing time before use on the baking plant. Alternatively, the dough may be salvaged by adding some flour and giving a **minimal** mix to disperse this extra flour. The dough will inevitably be overmixed and this will probably affect the quality of the biscuits made from it (see Fig. 10).

The best quality biscuits will always be made from dough that has been made by at least a two stage mixing process (see Section 5.3) and has had a minimum of mixing after the flour has been added. If there has been too much mixing the usual results are that the biscuits

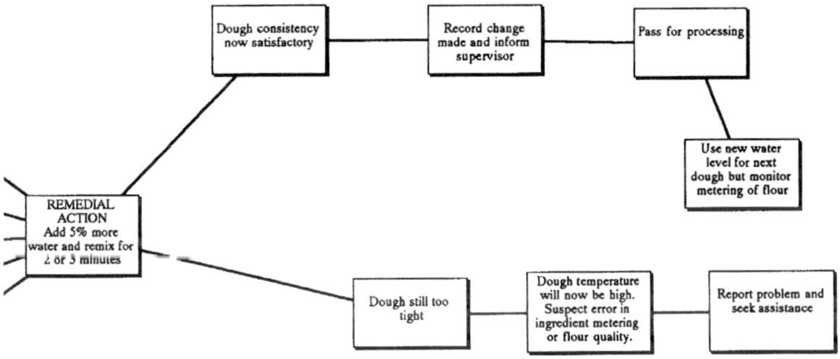

do not develop well in the oven or may show shrinkage in one axis after baking. This is because too much gluten development has occurred. This type of problem is likely to be seen most in short doughs with relatively low fat levels.

The converse problem is sometimes encountered. Namely that there is too little mixing at the final stage. The result is that the dough is very short and will not make a cohesive sheet or will not extract satisfactorily from a rotary moulder. Also, during baking there may be too much spread and not enough thickness development. A **little** more mixing at the final stage will help the dough to become more cohesive. This type of problem is likely to be seen most in short doughs with relatively high fat levels where the water level is low.

If the **fat is hard** (usually because it is too cold) it will be difficult to disperse and emulsify during the first stage of the mixing. If lumps persist after the 'cream up' mixing stage and are only broken up during the final stage of the mixing it is probable that the dough

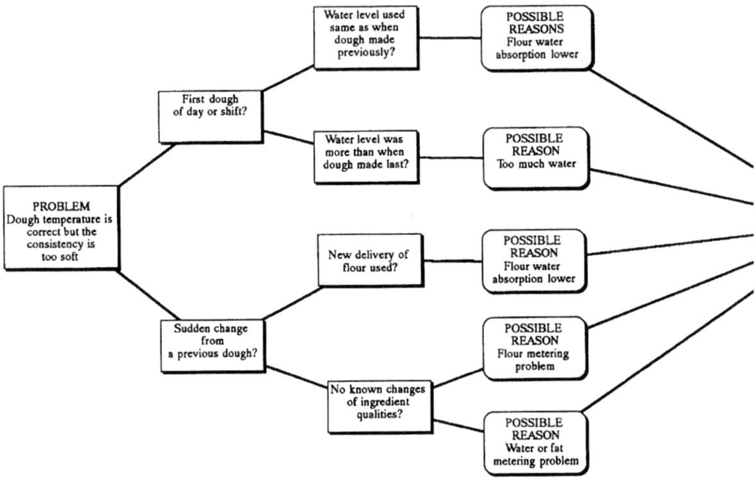

8 Problem solving – developed dough that is too soft at the end of the mixing programme.

quality will be unsatisfactory. A smooth, creamy and homogeneous mixture should be aimed for before adding the flour.

Fat that is hard can be 'improved' by plasticising through a machine that works it but the best solution is to store the fat at a specific temperature so that it is soft at the time it is needed. Unfortunately, warming a block of cold fat takes a long time. If the fat is heated quickly so that part or all of it becomes liquid the quality of a dough made with it will not be satisfactory.

The mixing procedure for short doughs means that the dough temperature is largely determined by the temperature of the ingredients. This means that in warm weather less water may be required to give the desired consistency.

7.3.3 Actions in the event of breaks in production delaying the requirement for dough

In order to achieve optimum quality of dough, mixings should be

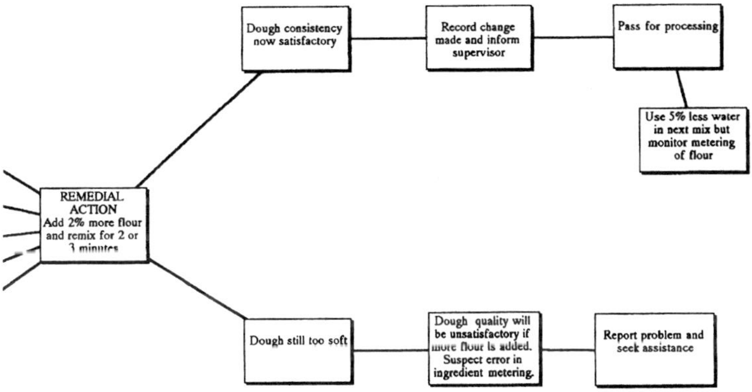

made according to a timetable that ensures that they are completed so that the standing times are similar for all batches.

Occasionally there is a break in production which is long enough to disrupt the requirement for dough. It is necessary that the mixing operator responds appropriately to these events.

7.3.3.1 Developed doughs

- Do not start the mixing process until the requirement for the dough is assured.
- If a dough has been partially mixed, interrupt the metering and mixing sequence before or during the final mixing stage and restart when the demand for the dough returns. The dough leaving the mixer will then be freshly mixed.
- If a dough has been mixed, retain it in the mixer so that a short remixing can be given when the demand for the dough returns. Ensure that the additional mixing does not take the temperature of the dough too high.

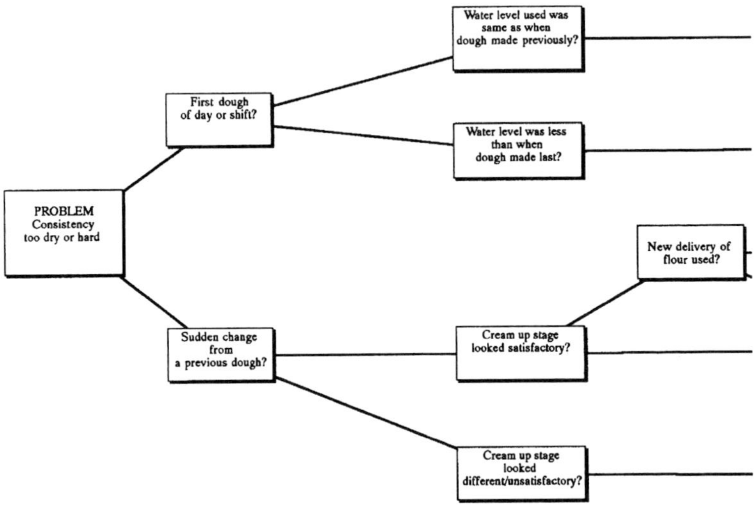

9 Problem solving – short dough that is too dry or too hard at the end of the mixing programme.

- If a dough has been mixed and ejected from the mixer, protect it from cooling and drying as best possible. A slightly longer standing period before use may be acceptable.

7.3.3.2 Short doughs

- Do not start the mixing process until the requirement for the dough is assured.
- If only the first stage of the mixing has been made, that is the cream up, stop the mixer and do not add the flour until the demand for the dough returns. It is satisfactory to let the cream up stage of short dough mixings wait for long times before the final stage is commenced.
- If the flour has been added, continue mixing to normal completion and eject the dough from the mixer. A longer standing time will probably be acceptable for most short doughs. Protect the dough from excessive heat or cooling.

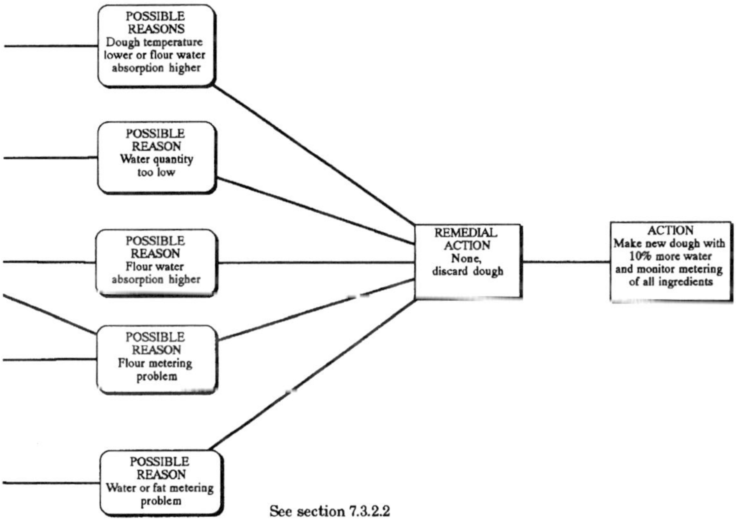

7.3.3.3 Fermented and enzyme modified doughs

The doughs must be mixed at intervals equal to the usage rate. The mixed doughs are placed in holding bins and stored for a specific time. It is not usual to modify the mixing programme unless the stoppage later in the process is seen to be very long. Refer to Sections 9.2.2 and 9.2.3.

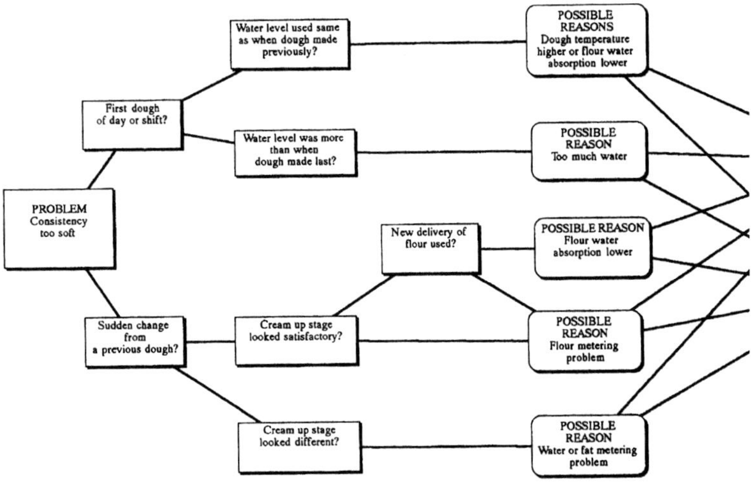

10 Problem solving – short dough that is too soft at the end of the mixing programme.

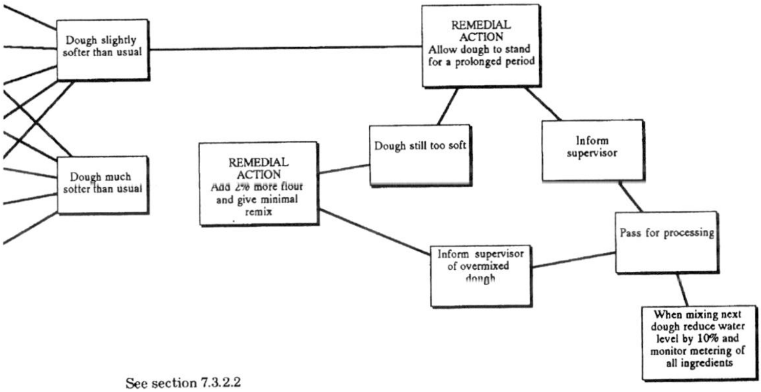

See section 7.3.2.2

8 Metering of ingredients into mixers

8.1 Introduction

A biscuit is made from a dough with a specific composition. This is known as the recipe or formula. Typically the dough is mixed in a batch mixer and the size of the recipe is adjusted to suit the mixing machine and the frequency with which the doughs are to be made.

The formula can be adjusted only to a limited extent otherwise the quality of the baked biscuit will be changed. These adjustments are typically limited to the water level (which affects the dough consistency) and the aerating chemicals (which affect the development of the biscuit thickness during baking).

Accurate weighing or metering of the ingredients for a dough is the key to much of the success of the biscuit making operation.

8.2 Handling ingredients

8.2.1 Manual weighing

In many factories some or all of the ingredients are carried to the mixer and weighed by hand. This is laborious and slow but with care is the most accurate and reliable method for assembling all the components of the dough. Weighing scales can be used that are appropriate to the size of the quantity to be weighed and each item can be ticked off as it is put into the mixer bowl. Those ingredients that must be sieved, to remove lumps, or dispersed in water before adding to the mixer can be dealt with easily and reliably.

The disadvantages of manual loading of a mixer include:

- Operations are slow and very labour intensive.
- It is hard work to lift large quantities of ingredients onto a weigher and then into the mixer bowl, especially if the mixer bowls are large and high off the ground.
- It is potentially unhygienic to tip ingredients from bags or boxes

as dirt that is picked up on the outsides of the containers may fall into the dough.

- There is a risk of contaminating the dough with string, plastic and paper from the containers of the ingredients.

8.2.2 Bulk handling of dry materials

In order to speed up the process of loading a mixer and also to reduce the hard work involved, many factories have introduced bulk handling systems for at least some of the ingredients needed for doughs.

Bulk handling means that the ingredient is conveyed from a large container (a silo) and metered as required directly to the mixer. There are usually several mixers fed by the system which is electronically sequenced and controlled. Dry materials such as flour and sugar are usually conveyed by a stream of air through pipes, known as pneumatic conveying. The feed into the air stream has to be controlled to avoid blockages and removal from the air is usually by means of a cyclone.

There are three common methods of weighing dry materials in a bulk handling system:

1 **Dedicated metering by weighing in**
The cyclone is an integral part of the weigher which is directly above the mixer where the material is to be used. When the weight is nearly made the flow of material from the silo is stopped and just before the weight is made the diverter valve into the weigh hopper closes. Any material remaining in the pipeline is returned to the silo from whence it came (in the case of flour) or is sent to another bin (for example, in the case of sugar). The precision of weighing is affected by the rate at which the material is conveyed because at the moment that the valve on the top of the weigher closes there will always be some material still in flight which will decide exactly how much arrives in the weigher. The rate at which the material is conveyed from the silo is set to suit the system and the number of calls per hour that will be made from all the mixers.

The materials are weighed for the next mix while the later stages of the mixing sequence are proceeding. They are then ready to be dropped by gravity directly into the mixer when required by opening a valve. There is sometimes a problem because not all of the material held in the weigher falls into the mixer. A visual display of the weight in the weigh hopper should be available for the mixer operator so that he can check that discharge has been complete.

2 **Central metering by weighing in**
The sequence is similar to that described above except that the weighing occurs in a centrally placed weigh hopper and this is discharged pneumatically to one of many mixers. The advantage of such a system is that there is less equipment to buy and maintain. The disadvantage is that it is slower to feed several mixers and there is a possibility that all of the flour leaving the weigh hopper does not reach the desired mixer because it is held in the pipe work, and what remains goes by mistake to the next mixer to call thereby causing a weighing error.

3 **Loss in weight metering (weighing out)**
This system involves a feed hopper above the mixer which is filled periodically from the silo in a non-critical manner. As required, material drops into the mixer beneath and the quantity is controlled by loss in weight.

Any number of silos or tanks can be arranged over the mixer, each mounted on load cells, and many ingredients can be metered into the mixer simultaneously. The loss in weight system affords great accuracy but is handicapped by the need and expense of a great amount of dedicated equipment. Mixers filled by loss in weight systems can be loaded much more quickly than weighing-in systems.

8.2.3 Bulk handling of liquid materials

Bulk handling of liquids and semi-solids (like plasticised fat) is somewhat easier than handling dry materials because they can be conveyed and metered by pumping.

Pipework can be connected directly to the top of the mixer. In the case of fats and syrups the pipework must be maintained at an appropriate temperature by jacketing.

Liquids can also be metered by the loss in weight system if they have been held in containers above the mixer. Such a system also lends itself to solutions of ingredients (premixes) that have to be prepared in relatively small quantities.

8.2.4 Weighing the mixer

This is a special case of the weighing in system. The whole mixer is mounted on load cells or a weigh frame and acts as the weigh hopper for material conveyed to it. This means that there is no error between the weighing operation and the dropping of that material weighed into the mixer.

The advantages of this system are that weight and time records can be kept of all ingredients delivered to the mixer whether added automatically or by hand. It is also possible to know how much dough was discharged from the mixer after completion of the mix. No other metering arrangements are required. However, the disadvantages are that separate holding bins and tanks are required above the mixer for each ingredient, similar to the arrangement for loss-in-weight. The weigh system must be robust to withstand the massive tare weight of the mixer and vibration when it is running yet sensitive enough to record precisely small ingredient additions. Operators must not touch the mixer while it is weighing. There is an 'in flight' problem as described above and only one ingredient can be metered at once. This prolongs the total time needed to load all the ingredients of a dough.

8.2.5 Continuous metering

A continuous mixer obviously needs to have all the ingredients of the dough supplied to it continuously and uniformly. This is a serious engineering challenge and is one of the reasons why this method of dough preparation is not commonly used. All the ingredients must be metered continuously and a combination of screw and belt feeders and pumps is used. By the use of loss in weight electronics the performance of these metering instruments can be controlled more precisely. It is usual to improve the metering of small ingredients by either diluting them with a filler like flour or dextrose or dissolving them to make a solution that can be more easily handled. To reduce the number of metering machines needed it is also common to make batches of premixes which combine a number of ingredients in the correct ratio.

8.3 Ingredients that should be dispersed before mixing commences

Most factories operate a partial bulk handling system for their ingredients. The reason for this is that it is cost effective to bulk handle the large ingredients but relatively expensive to mechanise the metering of small ingredients which often require special attention before they are added to the mixer.

In order to speed up the metering and to maintain accuracy most manufacturers dispense the small ingredients like salt, baking chemicals and flavours into bags or containers enough for all the mixings of a shift. The mixing operator then has merely to use one container of each for each mixing. It is helpful to use different coloured containers for the various ingredients to aid identification and to reduce mistakes.

During mixing there is usually good dispersion of even the smallest ingredients. However there are certain ingredients that tend to lump either during storage or when mixed in a dough. If these lumps remain biscuits may show black spots or have an uneven structure after baking.

It is advisable to sieve sodium bicarbonate before use, to dissolve ammonium bicarbonate and to disperse milk and whey powders in water with a high sheer action.

Sodium metabisulphite (SMS) is normally used as a 10% solution to aid metering of the very small quantities used.

Emulsifiers, like lecithin, work most effectively if they are well dispersed in the fat before an emulsion with water is made. If blocks of fat and emulsifier are used for a mixing it is worth considering making a preliminary mixing of the fat, emulsifier and possibly the sugar, before the other ingredients are added.

Fresh yeast should be dispersed in some water and when adding to the mix care should be taken to avoid putting the suspension near to the salt or other chemicals as these may kill the yeast cells.

Proteinase enzyme, which may be used in savoury cracker doughs, must be handled with care especially if it is a powder. Place the enzyme into the mixer so that there is no splashing or dust. The dust may cause irritation and damage if inhaled.

If small quantities of liquids are to be handled remember to use only plastic or metal containers. Never use glass in the factory environment.

The action of adding small ingredients to a mix requires some care and attention. Just in case the blending action of the mixer is not ideal or if there is only a short mixing time after the material has been added, it is best to spread the ingredient across the width of the mixer bowl.

8.4. Communication and recording changes

Ingredient metering errors are the major source of biscuit processing problems. Typically they cause much more quality problems than variations in ingredient quality.

The mixing operator is the most likely person to notice irregularities in the metering systems and he or she may have to be the person who makes corrections. It is very important that management and maintenance engineers have full details of the nature and occurrence of errors so that remedial action can be taken. It is therefore essential that good and full records are kept of the metering problems and what action was taken.

8.5 Troubleshooting

Several of the potential problem areas have been mentioned in the descriptions above.

8.5.1 Failure of an ingredient to make the correct weight

There is usually a visual indication of the weights of dry ingredients like flour and sugar that are being metered to the mixer. If insufficient material is metered or is delivered to the weigher by the bulk handling system the mixer sequence will wait and will not proceed (see also Fig. 6, p 40).

If it is noticed that there is a shortage of an ingredient so that the sequence will not proceed, and that it is not correcting itself, it is best to pause the sequence, to add the required amount of material by hand and then to proceed through the rest of the programme. This can be done either manually or if possible by setting instructions for the process to continue from where it was paused. There is a great danger that if the sequence is cancelled, a restart will mean that the process starts from the beginning and not from the point it had reached when the problem was discovered.

If the same problem is encountered in subsequent mixes engineering assistance should be called to reset the metering mechanisms.

9 Handling dough after mixing

9.1 Removal from the mixer

When mixing of a batch of dough is complete it is normal to eject it immediately either into a tub/trough or directly into a feed hopper. This leaves the mixer free for a new mix. If, for any reason, it is desired to retain the dough in the mixer bowl for a period and the mixer has a water jacket around the bowl, ensure that the temperature of this jacket is similar to that of the temperature of the freshly mixed dough. If it is higher or lower there will be a local change in temperature of the dough which will mean that the dough will have a variable consistency when it is being machined into dough pieces.

In the case of vertical spindle mixers the mixing bowl is the tub which is used to take away the dough. In other cases after the end of mixing the mixer is opened or the bowl is tilted to remove the dough. In many cases the ejection of the dough is not completely automatic and some 'help' is required. This should be done with care to ensure that no dirt or other materials become included in the dough. Use only metal tools to aid in the movement of the dough and do not hit the beaters or sides of the mixer bowl otherwise metal fragments may be lost into the dough.

Aim to empty the mixer completely, any dough that remains will behave like scrap dough and may affect the quality of the dough in the next mixing.

The dough will be used either without delay, as in the case of semi-sweet and some cracker doughs, or after standing for a specified period.

9.2 Standing the dough

The period of time and the temperature conditions while dough is stood prior to use will be critical for all types of dough. The main types will be described.

9.2.1 Semi-sweet dough

It is considered best to mix these doughs to a relatively high and specified temperature (for example 41°C). After removal from the mixer, these doughs begin to cool and dry at the surface eventually becoming hard and crusty. As a result of the mixing action and the heat which is developed, the gluten becomes extensible (it can be pulled out without breaking). As the dough stands and/or cools the extensibility and consistency of the dough deteriorate fairly rapidly. It is therefore desirable to use the dough without delay and at all times to protect it from cooling and drying. The size of each mixing of dough should be as small as possible compatible with the ingredient metering and mixing cycle time needed to produce a new dough and the rate that the dough is used.

9.2.2 Fermented cracker dough (cream crackers and soda crackers)

Yeast is used to modify the quality of many cracker doughs by a process known as fermentation. The yeast cells grow and multiply feeding on components in the warm dough. The products of fermentation are alcohol, which gives a recognisable smell, and carbon dioxide gas which is held in the dough and causes it to rise in the tub. If the fermentation period is longer than 8 h other changes will also occur in the dough as a result of bacterial growth. These changes include acid production and flavour change which contribute to the quality of the baked cracker.

Cracker doughs are relatively wet and must be protected from drying. The action (growth) of microorganisms is affected considerably by the temperature of the dough. It is therefore critical to both mix the dough to a specified final temperature and to store it at a similar temperature so that the fermentation action is uniform within the dough and from one dough to the next.

It is normal to ferment cracker doughs in rooms that have controlled humidity (of 80–90% RH, relative humidity) and temperature. Great care must be taken not to leave doors open as it takes some time for the humidity and temperature conditions to return to the desired levels if they are allowed to fall. If the storage area does not have humidity control, close fitting lids must be placed over the tubs.

Fermentation procedures vary from product to product and from factory to factory. Sometimes there is a remix stage which may involve additions of other ingredients.

9.2.3 Enzyme modified cracker doughs

Another type of cracker dough requires the action of an enzyme which alters the nature of the gluten. The enzyme is a proteinase and is added during mixing as either a powder or a liquid. The amount of the action is dependent on time, the quantity of the enzyme that is used and the temperature. The temperature/time/enzyme quantity relationship is specified relative to the time taken to process a batch of dough and the amount of enzyme modification required for a particular product. If the speed of reaction is too fast there may be a significant difference between the biscuits made at the start of a dough batch compared with those made at the end. If the reaction is too slow there will be many tubs of dough waiting on the factory floor and mixing the first dough of a production run may be at an inconvenient time.

It is essential that the temperature of the mixed dough is close to the specified value and the time that the dough stands is precisely controlled. As is the case for fermenting doughs, humidity control is required to prevent surface drying.

9.2.4 Puff doughs

Puff doughs are a special type of developed dough. In the course of dough piece forming a laminated structure must be developed and the layers are separated by soft, semi-solid, plasticised fat. The fat must be kept cool which means that the dough must also be cool or cold. Thus, puff doughs must be mixed and must stand so as to maintain low temperatures (usually around 15°C or less). Iced water is used at dough mixing and a chilled room is required to hold the dough and in which to make the laminations.

9.2.5 Short doughs

The stage when flour is incorporated into a short dough is typically of short duration. This ensures that little or no gluten is formed. The dough is short, the opposite of extensible: it cannot be pulled out without breaking. There is a period after mixing when water is absorbed by the flour and any other cereal ingredients. The dough thus appears to be drying and becomes much less sticky. It also becomes harder. These changes soon stabilise and thereafter little gluten can be formed if the dough is mixed or worked in dough piece forming machinery.

The combination of the need to have little gluten in a soft dough and yet a more or less uniform consistency of dough while a batch of dough is being formed into dough pieces, places special requirements on the handling of short dough.

It can be shown that the change in consistency due to water being absorbed becomes slower and slower. The maximum effect is in the first 30 min or so, thereafter the change is less noticeable although it does continue for several hours. It is therefore recommended that all short doughs are allowed to stand, without any agitation, for at least 30 min after completion of mixing.

If the holding period for short dough is in the hopper of a sheeter or pre-sheeter, make sure that the dough is not 'mixed' by the action of any agitators such as those that are designed to prevent bridging of the dough over the rollers. 'Mixing' by agitators may result in a dough that is tough, an effect similar to that caused by over mixing.

The consistency of a short dough may be much softer at the end of mixing than can be tolerated by the rotary moulder or other dough piece forming machinery owing to changes that occur during standing. Some experience is therefore needed to estimate the optimum consistency of a dough at the end of the mixing time.

There are some sugar and syrup-rich short doughs that are exceedingly sticky after mixing. The stickiness is reduced during the standing period. If this standing period is very long even the stickiest doughs become manageable in a sheeter or rotary moulder.

9.3 Removal of dough from the tub

Dough contained in a tub or trough is taken to the appropriate hopper and is transferred normally by tipping but sometimes by manual transfer. There are various mechanisms for tipping. In each case great care must be taken to ensure that dirt does not fall into the dough from external parts of the tub, such as its wheels or the forks of a tub tilter.

Empty tubs should always be covered to prevent dirt collecting in them prior to them being used for more fresh dough.

9.4 Troubleshooting

The most likely problems that are encountered with dough handling are as the result of **breaks in production** such that the dough stands for longer than normal. Action must be considered carefully.

9.4.1 Semi-sweet and other developed doughs

These doughs deteriorate in quality with standing especially if they are allowed to cool. An old dough will become shorter, less extensible, and this will affect the shape and other qualities of biscuits made from it. These doughs should not be used if they are more than twice as old as normal. Some improvement can however be achieved if the dough is remixed but in this case the temperature should not be allowed to rise significantly above the normal value.

9.4.2 Fermented and enzyme-modified doughs

Time is critical for the quality of these doughs. Because there will be several doughs standing, in the event of a delay in usage it is normally best to discard any doughs that have stood for more than 15% longer than the normal time. By discarding one or more doughs it is possible to return to the normal dough standing time. If batches of dough are not discarded there will be a prolonged period of production using overfermented dough.

9.4.3 Short doughs

In contrast to developed and fermented doughs extra standing time will not be so detrimental to short doughs. They do slowly become firmer in consistency but where a normal standing time is about 30 min a standing of 90 min will probably still result in satisfactory dough.

10 Care, maintenance and cleaning of machinery

10.1 Care

The equipment in the mixing department looks very robust because the mixing machines and other equipment are mostly large and made of metal. While this is generally true, operatives in this area should exercise care, not only to avoid hurting themselves and their fellow workers but also to make sure that there are no interruptions in the supply of dough to the subsequent production department.

The main aspects of careful behaviour are described in Section 3, *Hygiene and safety aspects*.

10.1.1 Consideration of others

- Tubs of dough are heavy. Special care should be taken when pushing these that they do not go out of control and hit someone.
- If water or fat is spilled onto the floor it should be cleaned up immediately because slippery floors can be an extra hazard when tubs have to be pushed and guided.
- If you are working at a high level, for example feeding an ingredient into the top of a machine, take extra care not to drop or spill anything on someone who may be standing or working below you.
- If you see a fellow worker in a dangerous situation give a warning. If someone is in trouble or struggling to handle something, give a hand.

10.1.2 Consideration of equipment

- Cleaning of equipment is everyone's responsibility. There should be preset tasks and methods, keep to them and do the jobs thoroughly.
- Avoiding spilling or spraying water on electrical components or lubricated bearings.

- Metering equipment and instruments are very sensitive and are easily damaged by mishandling. Do not hammer or kick them!
- If difficulties or malfunctioning are encountered attend to the problem thoughtfully or seek assistance. If the malfunctions occur repeatedly report them so that engineering attention can be given before something breaks down completely.

10.1.3 Consideration of possible contamination of the dough

- Assembling ingredients into a mixer and handling of dough offer many opportunities for contamination. *Sources of contamination* are described in full in Section 3.2.
- Cover dough in tubs, especially if it is not to be used for a while or is scrap dough that will be used on another occasion.
- Of particular relevance in the mixing department are dangers of floor dirt from footwear if operators have to climb onto machinery to attend to problems or during cleaning operations.
- Pushing heavy dough tubs often results in collisions with walls or building pillars which thereby can become damaged and release plaster, tile chips or other masonry. This may fall into the dough. It is recommended that all walls where collision is possible should be shielded with steel plates.
- Do not hit metal surfaces with spades or long handled tools as slivers of metal may be released into the dough.
- Make sure that all hand tools such as handles, spanners and scrapers are kept in a safe place where they cannot fall into dough tubs either empty or containing dough. Dough tubs are not suitable containers for tools or cleaning materials.
- Insect and rodent infestation is a particularly serious source of contamination especially in the mixing department. Precautions should be in place to reduce the chance of insects, birds and animals entering a food factory but if they are seen report the matter to those responsible for controlling pests and cleaning.

10.2 Cleaning of mixers

- Cleaning mixers either at the end of a shift or before a different type of dough is made is the responsibility of mixing department operators.
- Mixers are powerful machines and very dangerous if not used according to the safety instructions. If it is necessary to reach or

climb into a mixer it is essential that the motor isolator switch is turned off. In most cases there is also a safety key that the person doing the cleaning should have in his possession before entering the mixer. This makes it impossible for the mixer to start accidentally or for someone else to start the machine.

- Observe great caution when using brushes which may have loose bristles and use only permitted detergents for cleaning the internal surfaces of mixers. Make sure that traces of detergents are washed away as they often have unpleasant smells and tastes if allowed to get into a dough.
- If, as the result of using water to clean mixing bowls, the beater bearings are seen to leak water, report the problem to engineering maintenance staff and clean up water that collects under the mixer.
- Clean under machines, paying particular attention to hidden areas under covers, etc. where pieces of dough may collect.
- Dispose of old pieces of dough and other unwanted materials by placing them in containers provided.
- Leave damp mixers open to allow ventilation and drying.

10.3 Maintenance

All machinery needs maintenance and repair from time to time. Doing repairs is not normally the job of mixing department operatives but your involvement in spotting and reporting problems is most useful to the efficient running of the factory. Look particularly for leaks of water, grease or dust. Report and record the problem precisely.

Useful reading and additional study

Almond N (1989) *Biscuits, Cookies and Crackers, Volume 2 The Biscuit Making Process*, Elsevier Science, London.

CABATEC (1989) *An Introduction to Biscuits*, An audio visual open learning module Ref. S8, The Biscuit, Cake, Chocolate and Confectionery Alliance, London.

CABATEC (1992) *Biscuit Mixing*, An audio visual open learning module Ref. S10, The Biscuit, Cake, Chocolate and Confectionery Alliance, London.

Manley D J R (1991) *Technology of Biscuits, Crackers and Cookies*, 2nd edition, Woodhead Publishing, Cambridge.

Index